UPDOG

WORLD'S GROSSEST

WORLD'S GROSSEST HUMAN BODY FACTS

Scott Nickel

Lerner Publications ◆ Minneapolis

Lerner Publications Company
An imprint of Lerner Publishing Group, Inc.
241 First Avenue North
Minneapolis, MN 55401 USA

For reading levels and more information, look up this title at www.lernerbooks.com.

Main body text set in ITC Franklin Gothic Std.
Typeface provided by Adobe Systems.

Designer: Mary Ross **Photo Editor:** Cynthia Zemlicka

Library of Congress Cataloging-in-Publication Data

Names: Nickel, Scott, author.
Title: World's grossest human body facts / Scott Nickel.
Description: Minneapolis, MN : Lerner Publications, 2022 | Series: World's grossest (updog books) | Includes bibliographical references and index. | Audience: Ages 8–11 | Audience: Grades K–1 | Summary: "From earwax and bad breath to boogers and burps, the human body is a disgusting place. Readers learn all about their bodies' sticky, smelly functions in this revolting tour of human anatomy"— Provided by publisher.
Identifiers: LCCN 2020012428 (print) | LCCN 2020012429 (ebook) | ISBN 9781728403984 (library binding) | ISBN 9781728418964 (ebook)
Subjects: LCSH: Human body—Juvenile literature. | Human physiology—Juvenile literature.
Classification: LCC QP37 .N565 2021 (print) | LCC QP37 (ebook) | DDC 612—dc23

LC record available at https://lccn.loc.gov/2020012428
LC ebook record available at https://lccn.loc.gov/2020012429

Manufactured in the United States of America
1-48456-48970-6/16/2021

TABLE OF CONTENTS

HEAD TO TOE

Our bodies are amazing—amazingly gross!

Dandruff is skin that flakes off from your scalp.

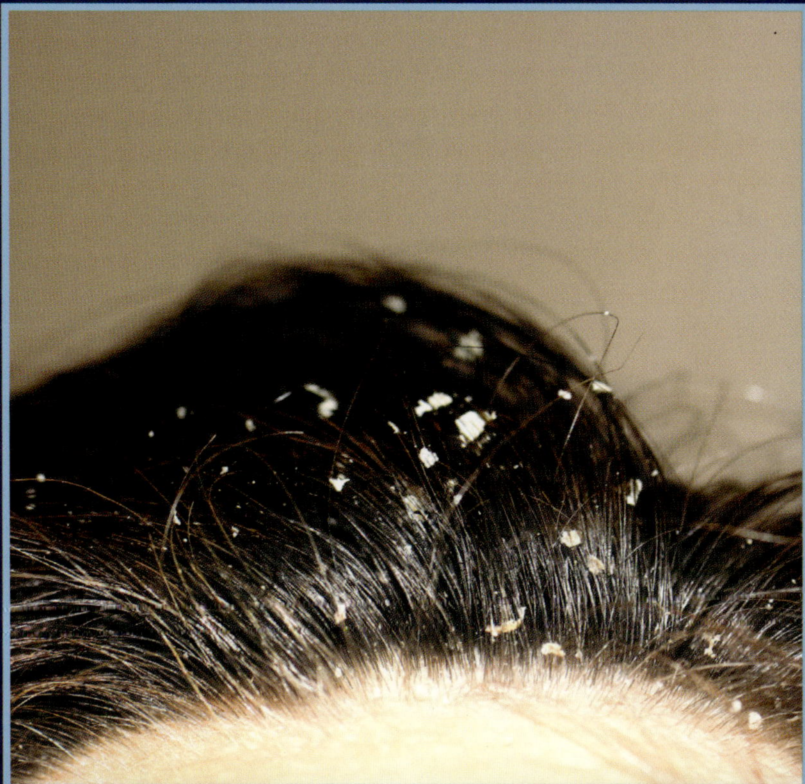

SCALP:
the skin on your head where hair grows

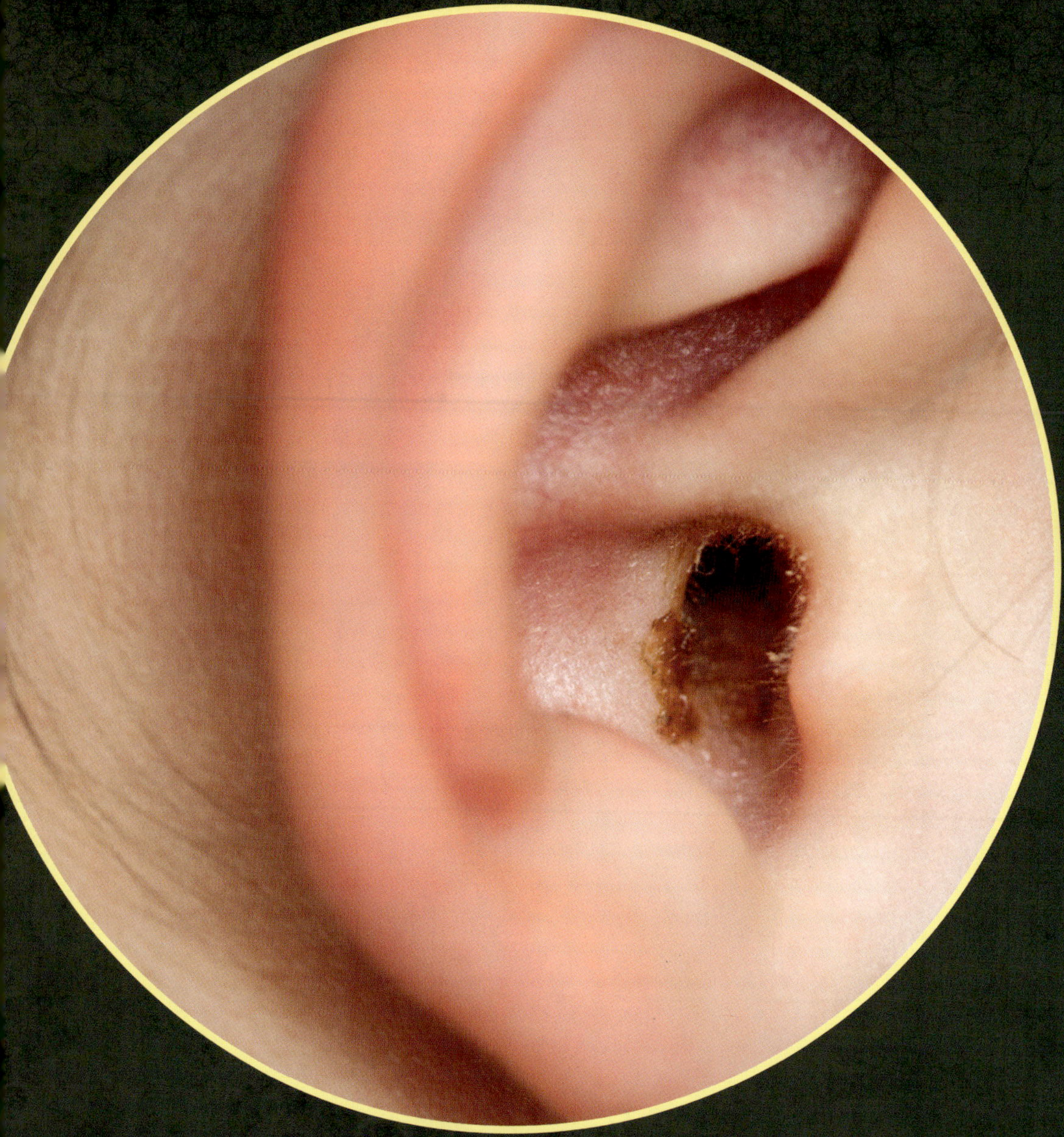

Earwax traps dust and germs so they don't hurt your ears.

Each of your feet can produce up to half a cup of sweat every day.

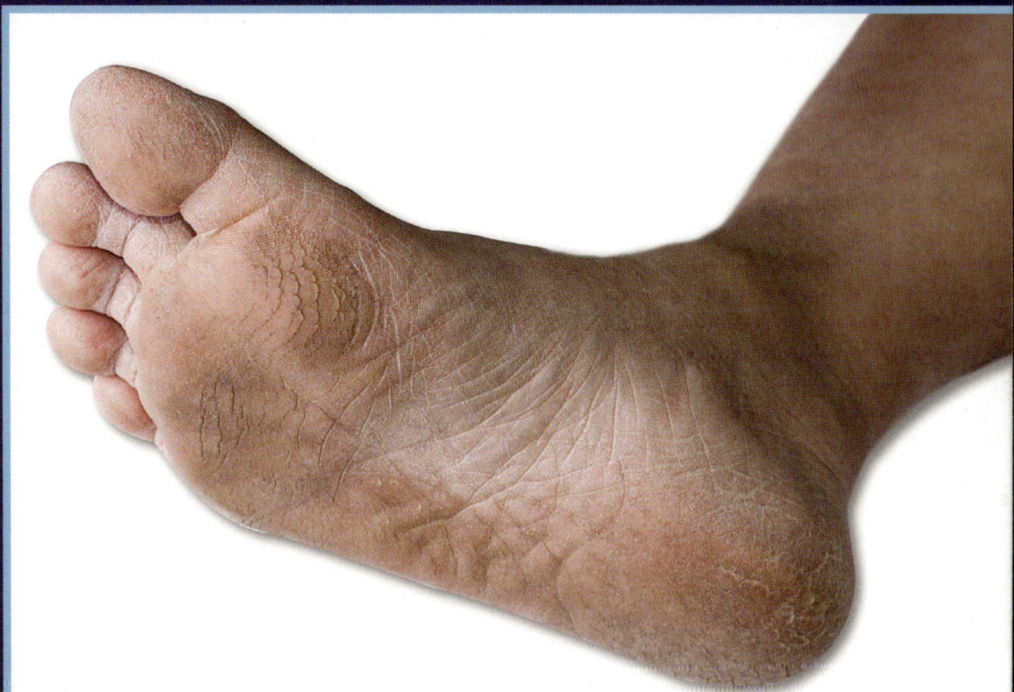

UP NEXT!

GROSS SOUNDS AND SMELLS.

FOUL BODILY FUNCTIONS

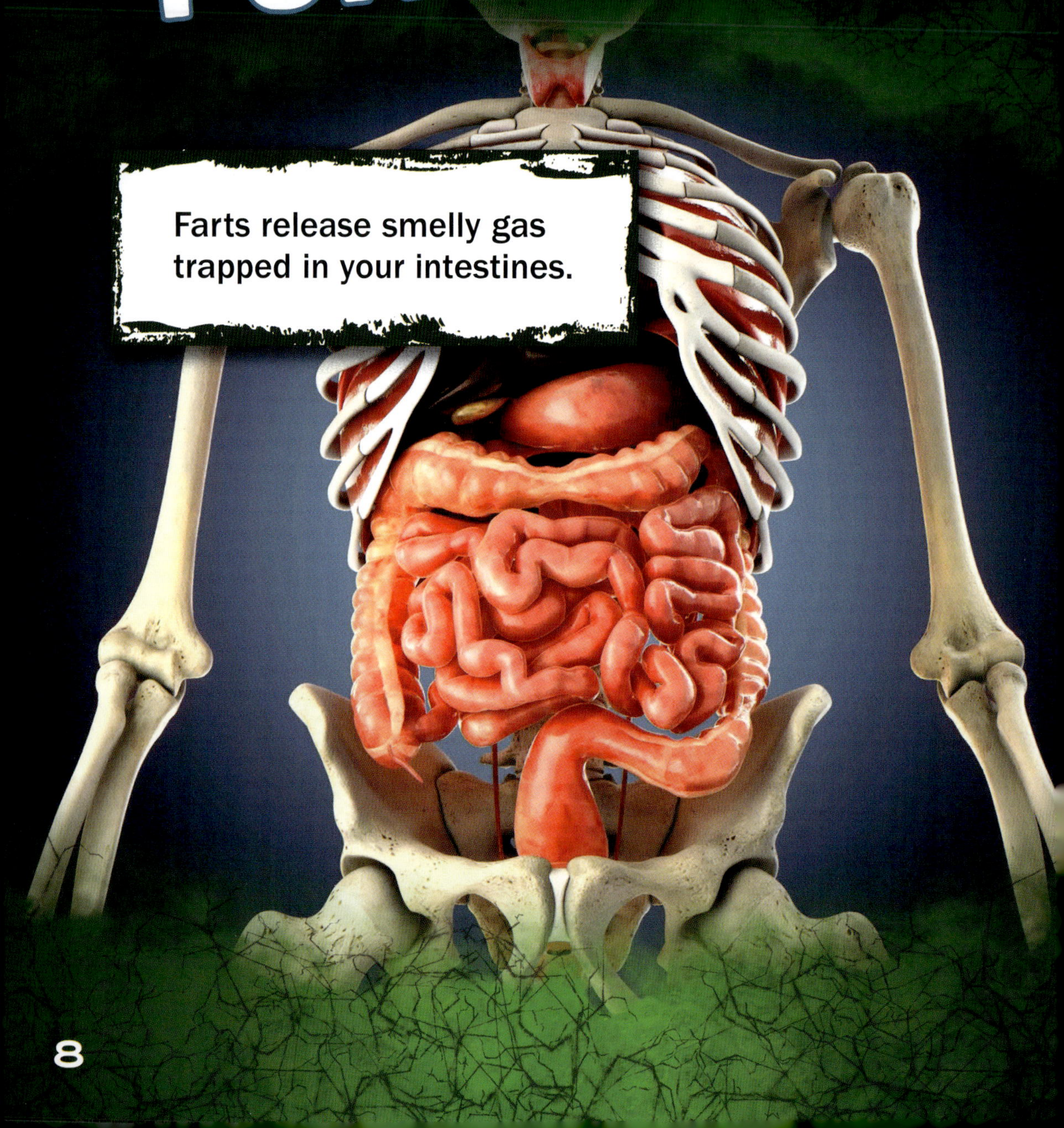

Farts release smelly gas trapped in your intestines.

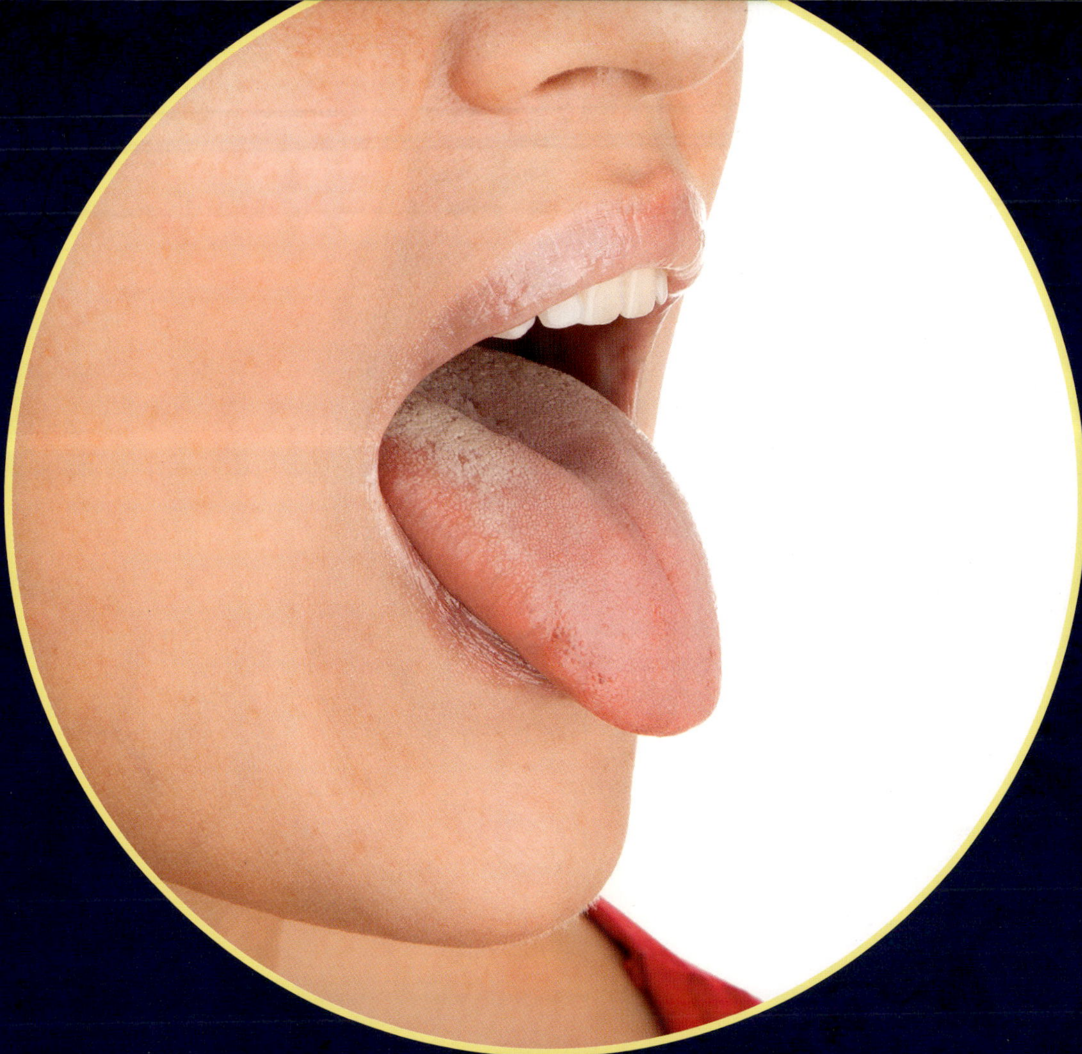

Drinking carbonated soda can give you burps.

CARBONATED:
fizzy or bubbly

WHICH IS GROSSER?

Snot vs. Pus

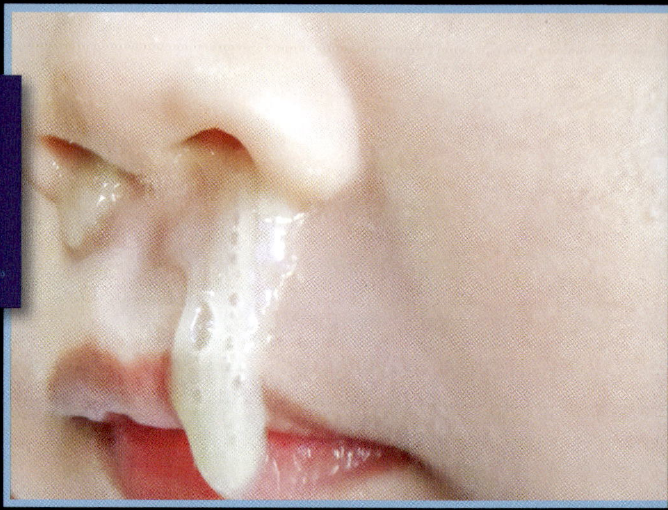

Snot is a sticky, slimy liquid called mucus. You swallow over two cups every day.

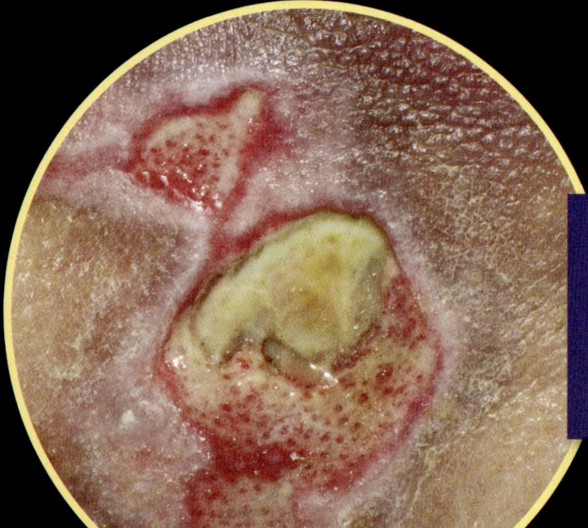

Pus is a yellow-green liquid around a cut. It means your body is fighting infection.

Bacteria in your mouth can cause smelly breath.

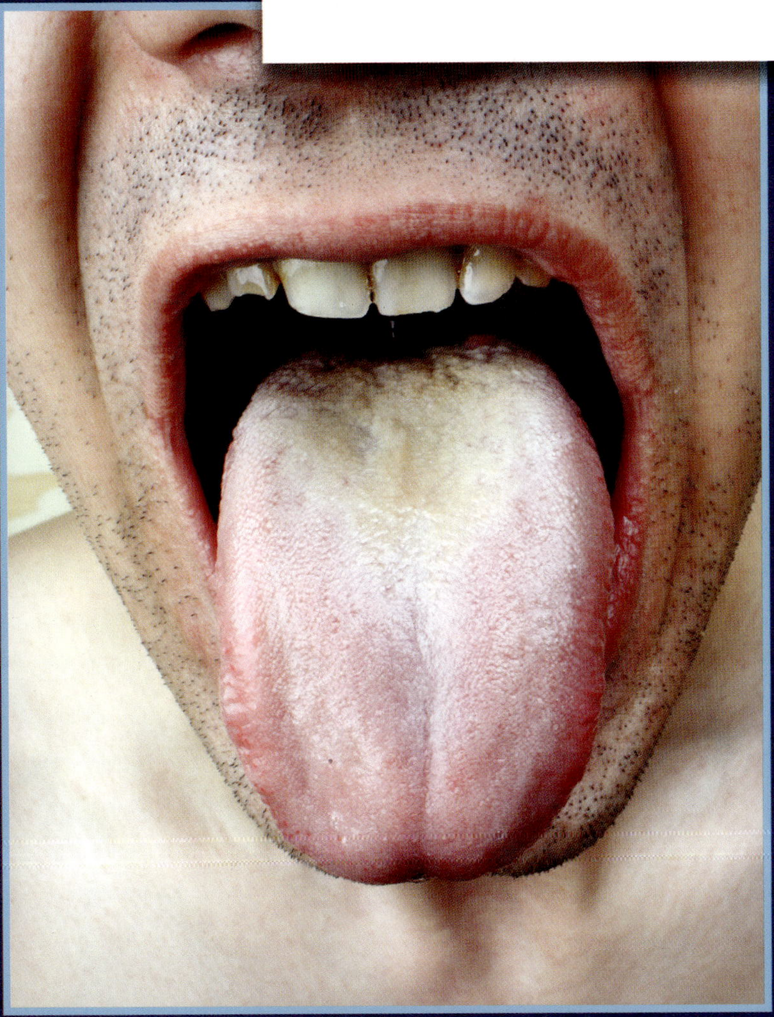

BACTERIA:
tiny living things, some of which can make you sick

DISGUSTING DIAGRAM!

Boogers are dried balls of snot.

Different snot colors have different meanings.

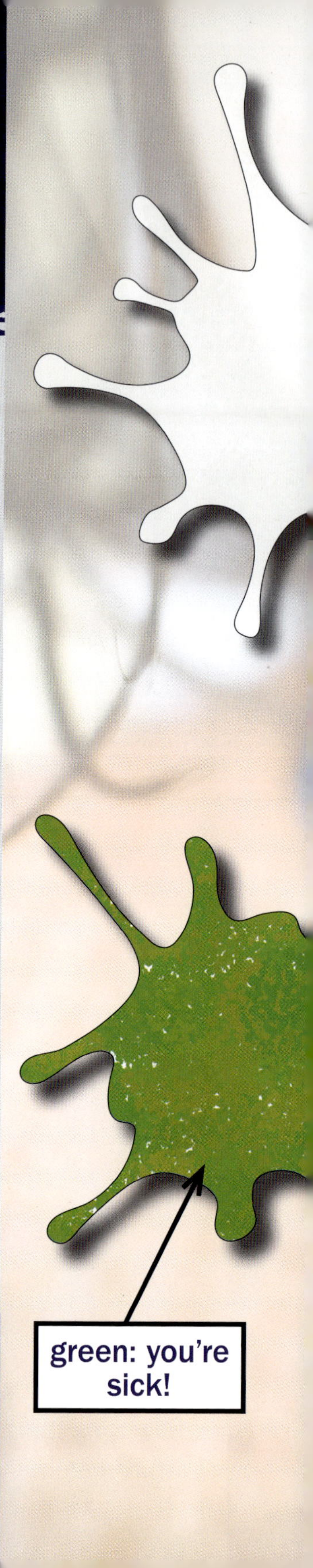

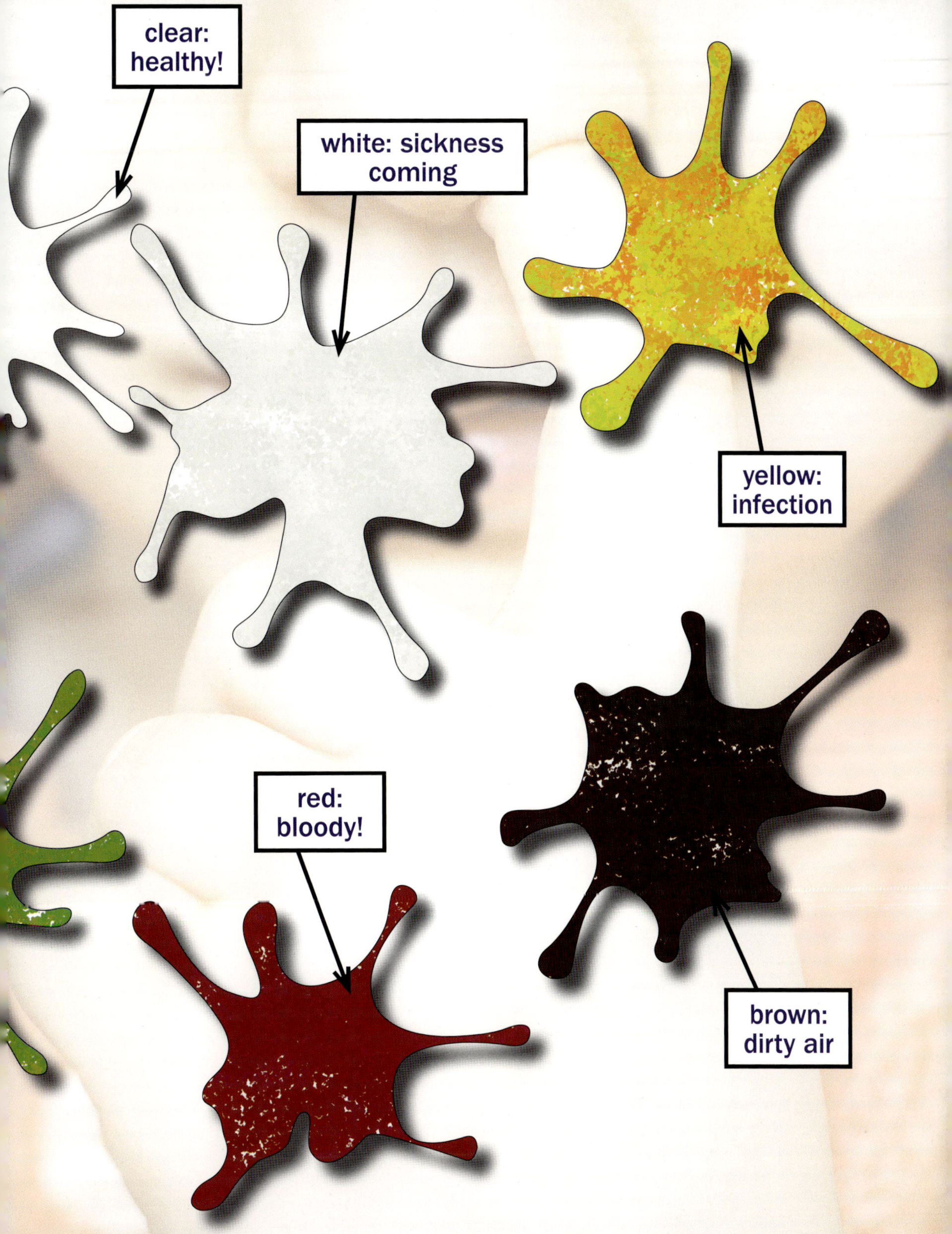
clear: healthy!
white: sickness coming
yellow: infection
red: bloody!
brown: dirty air

Peeing clears waste out of your body.

Diarrhea is poop that has too much water.

Spit is a mix of water and mucus.
Your body makes over a quart a day.

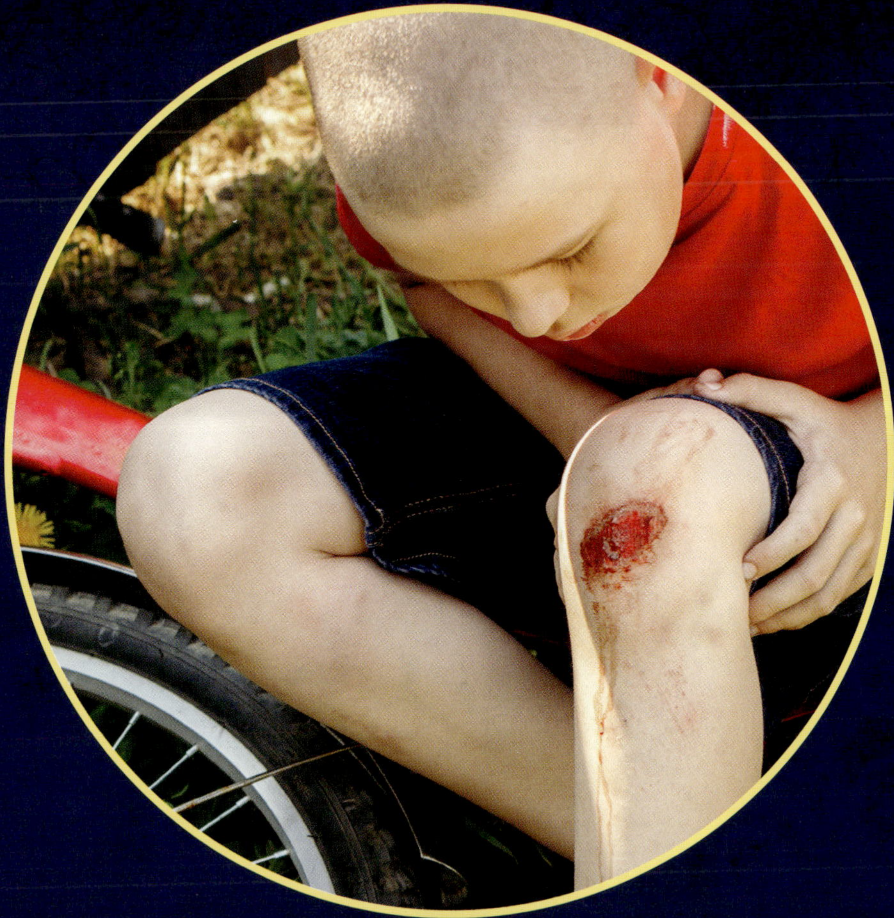

Your body is full of blood. An average man's body holds about a gallon and a half.

UP NEXT!

ITCHY AND ICKY.

SCARY SKIN

Rashes are red, itchy, and bumpy.

Scabs are hardened blood.
If they're infected, they can ooze.

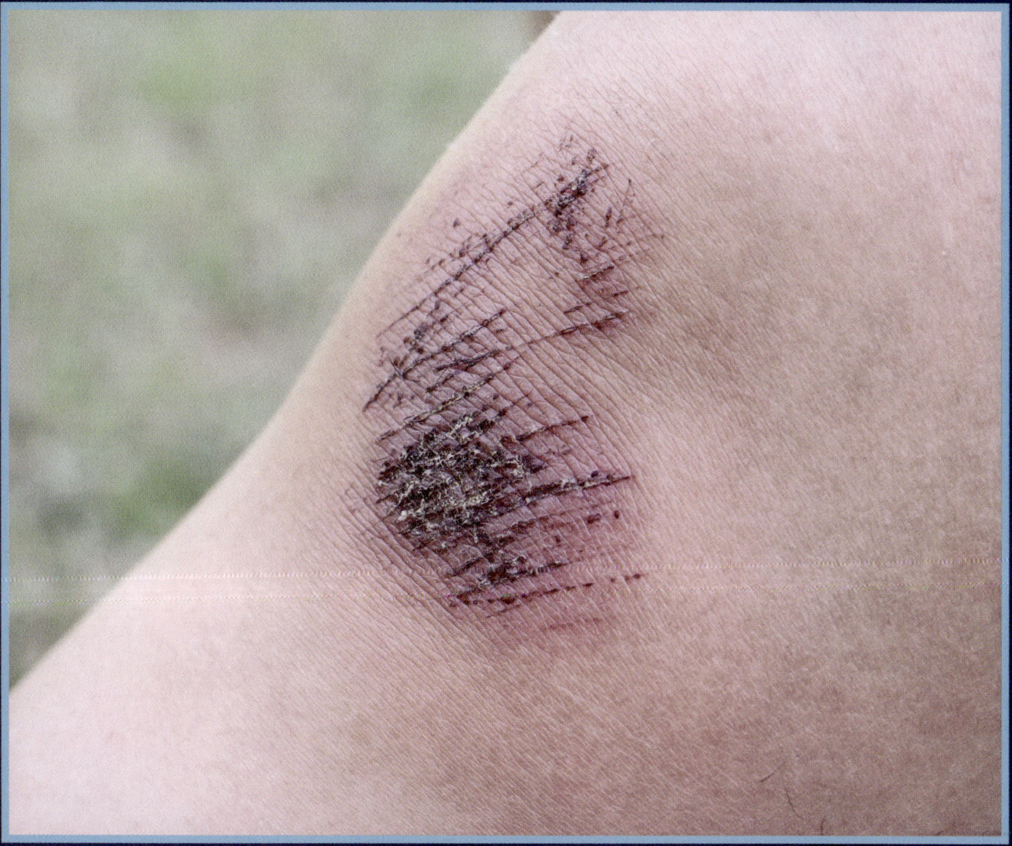

Bruises turn black and blue. They're damaged blood cells under your skin.

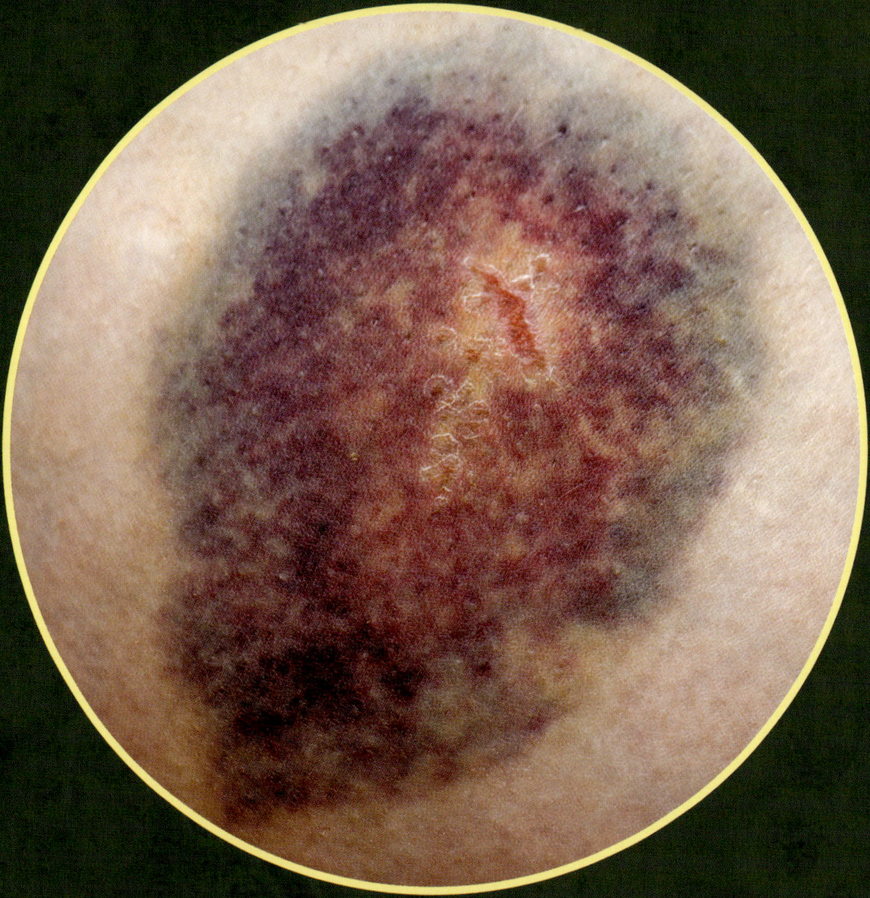

CELL:
any one of the tiny parts that together form all living things

From head to toe, our bodies are pretty gross!

Glossary

bacteria: tiny living things, some of which can make you sick

carbonated: fizzy or bubbly

cell: any one of the tiny parts that together form all living things

scalp: the skin on your head where hair grows

Check It Out!

Braun, Eric. *Awesome, Disgusting, Unusual Facts about the Human Body.* Mankato, MN: Black Rabbit Books, 2018.

Garstecki, Julia. *Pick, Spit & Scratch: The Science of Disgusting Habits.* Minneapolis: Quarto, 2018.

Gross Facts about the Human Body
https://sciencewithkids.com/science-facts/gross-body-facts.html

Human Body Facts for Kids
https://sciencewithkids.com/science-facts/human-body-facts-for-kids.html

National Geographic Kids
https://www.natgeokids.com/nz/discover/science/general-science/15-facts-about-the-human-body/

Nickel, Scott. *World's Grossest Jobs.* Minneapolis: Lerner Publications, 2022.

Index

Photo Acknowledgments

Image credits: Veres Production/Shutterstock.com, p. 4; Timonina/Shutterstock.com, p. 5; Andrii Spy_k/Shutterstock.com, pp. 6, 20; Kritsana Noisakul/Shutterstock.com, p. 7; u3d/Shutterstock.com, p. 8; fortton/Shutterstock.com, p. 9; Blur Life 1975/Shutterstock.com, p. 10 (top); Ciolanescu/Shutterstock.com, p. 10 (lower); Adam J/Shutterstock.com, p. 11; Denklim/Shutterstock.com, p. 13; RapidEye/Getty Images, p. 14; Daisy Daisy/Shutterstock.com, p. 15; David Trood/Getty Images, p. 16; ikrolevetc/Shutterstock.com, p. 17; TY Lim/Shutterstock.com, p. 18; Thanyaluck Namjuck/Shutterstock.com, p. 19; Shift Drive/Shutterstock.com, p. 21. Design elements: Saravudh Sudkhaw/Shutterstock.com; Green angel/Shutterstock.com.

Cover: Nerthuz/Shutterstock.com.